I0845831

Coded or Cultivated

A Journey through the Nature-Nurture Nexus

Freudian Trips

Copyright Page

© 2023 by Freudian Trips

All rights reserved. No part of this book may be reproduced in any form or by any electronic or mechanical means, including information storage and retrieval systems, without permission in writing from the publisher, except by a reviewer who may quote brief passages in a review.

This book is a work of non-fiction. Unless otherwise noted, the author and the publisher make no explicit guarantees as to the accuracy of the information contained in this book and will not be held responsible for any errors or omissions.

Published by Omniterra Media Inc

First Edition

Visit the author's website at www.freudiantrips.com

Disclaimer

The views and opinions expressed in this book are those of the author(s) and do not necessarily reflect the official policy or position of any other agency, organization, employer, or company. The contents of this book are for informational and educational purposes only and are not intended to serve as professional advice, diagnosis, or treatment.

The information provided in this book is believed to be accurate and reliable as of the date of publication. However, it may include some errors or inaccuracies, and no warranty or guarantee is provided regarding the accuracy, timeliness, or applicability of the content.

Readers are encouraged to consult with professional philosophers, educators, or other qualified professionals where appropriate for personalized advice. The author(s) and publisher shall not be liable for any loss, damage, or harm caused or alleged to be caused, directly or indirectly, by the information or ideas contained, suggested, or referenced in this book.

By reading this book, the reader acknowledges and agrees that they are solely responsible for how they interpret and apply the information contained herein.

This book may also include references to other works, studies, and sources. These references are provided for further reading and exploration and do not imply endorsement or validation of the specific theories, viewpoints, or interpretations presented in those works.

Chapter 1: Introduction to the Great Debate

Have you ever watched two toddlers play? One might quietly sit, engrossed in coloring a picture, while the other might be energetically racing toy cars across the room. They seem to be so different. Why? Is it because they were born that way or because of how they were raised?

Welcome to one of the oldest questions in the realm of understanding ourselves: the great debate of nature vs. nurture.

Imagine a time before all our technological advancements, even before we knew anything about DNA. Picture philosophers, thinkers, and ordinary folks debating around candlelit tables: Are we a blank slate when born, waiting for life's experiences to shape us? Or are we pre-wired with our destiny?

Nature

In this corner of the debate, we have the idea of "nature." Think of it as your body's rulebook, a manual that guides how you grow and develop. This rulebook is full of tiny instructions that determine everything from your eye color to, some argue, aspects of your personality or talent. These instructions come from our parents and their parents before them, all the way down the line. They're a part of our DNA. So when we talk about "nature," we're really talking about the genes and hereditary factors that contribute to our being.

Nurture

On the opposing side, we have "nurture." Picture this as the vast world around you: your parents, that favorite teacher you had in third grade, the movies you watch, the books you read, the friends you make, even the food you eat. All these experiences and interactions with the environment can shape you. Some argue that if you were raised in a different part of the world, with a different family or in a different decade, you might be a completely different person today. That's the power of nurture.

Why Does This Debate Matter?

This isn't just a fun debate for philosophers and scientists; it's a question that affects all of us. It delves deep into the core of who we are. Are we simply products of our genes, or can we change, adapt, and grow based on our experiences?

The nature vs. nurture debate touches everything from how we raise our children, approach education, deal with crime, and even how we treat illnesses. Think about it. If someone is naturally predisposed to a certain behavior or illness, should they be treated the same way as someone who might have developed it because of their environment? It's a tough question, and as you'll see in the coming chapters, the answer is not as straightforward as picking one side.

So, the next time you're marveling at your own quirks, talents, fears, or dreams, take a moment to ponder: Is it nature, nurture, or a little bit of both that's crafted the masterpiece that is you? Stick with us on this journey as we unravel the threads of this age-old debate, weaving through history, science, and human stories to find the answers.

Let's embark on this exploration together. Through the pages ahead, you're about to dive deep into a world of discovery that brings together tales of identical twins separated at birth, groundbreaking scientific experiments, and fascinating real-life anecdotes. So buckle up; it's going to be a thought-provoking ride!

Chapter 2: Nature's Impression

Imagine for a moment that you're assembling a jigsaw puzzle. Each piece has its unique shape, color, and design that fits perfectly into a specific spot. In many ways, our genes are like those puzzle pieces. They help piece together the vast and complex jigsaw puzzle that is you. In this chapter, we'll dive deep into the fascinating world of genetics and explore just how crucial these tiny puzzle pieces are in shaping who we are.

Our Genetic Blueprint

Let's begin with a basic idea: every human being starts as a single cell. And within this cell is a blueprint that holds instructions on how to build every part of you. These instructions are contained in our DNA, which is packed within our genes. Imagine DNA as a recipe book, and genes as individual recipes within that book. Each recipe determines various things about us, like our height, eye color, or even how curly or straight our hair is.

Twins: Nature's Perfect Experiment

One of the most compelling ways to see the power of genetics is by looking at identical twins. These twins come from a single fertilized egg and share 100% of their genes. But what happens if they grow up apart, in different families or even different countries?

There have been real-life instances where identical twins were separated at birth and raised in completely different environments. In many of these cases, when the twins reunite years later, the similarities are astonishing! They might have similar hobbies, tastes in food, ways of laughing, or even the same type of dog. These striking resemblances underscore the powerful role genes play in shaping who we are.

Genes and Health: A Double-Edged Sword

However, our genes don't just influence our personalities or looks; they can also affect our health. Some of us, due to our genetic code, might have a higher likelihood of developing certain diseases. For instance, some families might notice that many of their members have high blood pressure or diabetes. This doesn't mean they're guaranteed to get these conditions, but their genetic makeup might make them more vulnerable.

But here's an exciting part! Knowing about these genetic predispositions can be empowering. It can guide us to take preventive steps, like adopting a healthier lifestyle or getting regular check-ups.

As we piece together the jigsaw puzzle of human development, it's evident that our genes play a starring role. They're like the first draft of a story. However, as any writer would tell you, the first draft is just the beginning. How the story unfolds depends on so many other factors.

So, while our genes set the stage, it's the interplay with our environment, experiences, and choices that truly define who we become. Stay with us, for in the next chapter, we'll dive into the nurturing world around us and discover its profound influence on our lives.

Chapter 3: The Nurture Hypothesis

Picture this: Two seeds of the same plant species are sown. One in a sunlit garden with rich soil and the other in a dimly lit room with scarce resources. While their genetic blueprint is identical, their growth, health, and vitality would differ vastly. This paints a vivid picture of the immense influence of environment or, in human terms, nurture.

The World Around Us

Every day, from the moment we open our eyes in the morning to when we close them at night, our surroundings interact with us. These interactions can shape our thoughts, feelings, and behaviors. This is the essence of the "nurture" part of our journey. The places we go, the people we meet, the experiences we have, all leave a lasting impression on us, much like footprints in the sand.

Bandura's Revelations

Let's take a moment to travel back to a classic study that shook the very foundations of psychology. Dr. Bandura, a prominent psychologist, introduced children to a fun-looking inflatable toy called the Bobo doll. He showed them a video where an adult either played nicely with the doll or treated it aggressively.

Guess what happened when these kids were later left alone with the doll? Many of them imitated the adult's behavior! Children who saw aggressive actions often displayed the same, while those who watched peaceful interactions were gentler. This study provided clear evidence that our behavior isn't just built-in but can be shaped by watching and learning from others.

Parenting, Culture, and the Colors of Life

If Bandura showcased learning from observation, the diverse cultures and parenting styles worldwide show the broader spectrum of environmental influence.

In some cultures, children might be taught the value of individual achievements and competitiveness. In others, community and collaboration might be emphasized. These teachings mold our values, beliefs, and outlooks.

Similarly, parenting plays a pivotal role. A child showered with praise might grow up to be confident, while one frequently criticized could become reserved or anxious.

Beyond Visible Changes: The Dance of Genes and Environment

Now, here's a mind-bender: our environment can also whisper to our genes and influence how they act. This phenomenon is known as "epigenetics."

Think of it like this: imagine our genes are keys on a piano. Epigenetics is like a pianist's hand, deciding which keys to press, when, and how softly or loudly. So, while the keys (genes) remain the same, the music (gene expression) can change based on the pianist's choices (environmental influences).

The world around us, our experiences, and interactions leave an indelible mark on who we are. They sculpt and mold, challenge and nurture, teaching and guiding us at every turn. It's like a dance, a beautiful ballet where the dancers (nature and nurture) flow in harmony, crafting the unique performance that is each of our lives.

As our journey continues, remember, it's not just about the genes we are born with but also the stories we collect and the memories we create.

Chapter 4: The Confluence of Nature and Nurture

Imagine you're at the beach, watching the waves meet the shore. The ocean brings its force and depth, while the sand offers its warmth and embrace. Together, they create a mesmerizing dance of ebb and flow. This beautiful meeting of water and land is much like the relationship between our genes (nature) and our environment (nurture). They don't exist in isolation; instead, they flow together, shaping the tapestry of our lives.

The Dance of Two Forces

At the heart of our exploration is the interactionist perspective. It's like looking at the world through a special pair of glasses that allows us to see both the genetic coding and the environment, not as two opposing forces, but as partners in a dance.

Here's a simple way to think about it: genes might provide potential, but the environment decides how (or even if) this potential is realized. It's like having a car (genes) – its make and model determine the maximum speed, fuel efficiency, and comfort. But how fast you drive, where you go, and how often you refuel (environment) can greatly affect your journey.

When Genes and Environment Chat

This ongoing conversation between our genes and our surroundings is known as the gene-environment interaction. It's a bit like a dialogue between two old friends: one might suggest an idea (genes) and the other might modify it, reject it, or build upon it (environment).

A well-known example is height. While our genes have a say in how tall we might become, factors like nutrition during our growing years can significantly influence the final outcome. Two people with a genetic potential to be tall might end up with different heights if one had adequate nutrition while growing up and the other didn't.

Real-life Rendezvous of Nature and Nurture

Let's delve into some more examples that paint this picture clearer:

Musical Talent: Someone might be born with a genetic predisposition to have excellent hand-eye coordination and a keen ear for musical tones. But if they never encounter a musical instrument or are discouraged from exploring music, that potential might never fully blossom.

Learning Languages: Some folks might have a genetic edge in picking up languages quicker. But, without exposure to various languages or the motivation to learn, this ability might remain untapped.

Stress Resilience: Genetically, certain individuals might be predisposed to handle stress better. However, if they grow up in highly traumatic environments without adequate support, their innate resilience might be overwhelmed.

As we weave through the intricate narrative of life, it becomes clear that we aren't just products of our genes or mere reflections of our environment. We are a beautiful blend, a symphony where every note and pause, crescendo and decrescendo, has its significance.

Our journey forward will uncover more layers of this relationship, dispelling myths, uncovering truths, and celebrating the dance of nature and nurture. So, as we turn the page, let's embrace the complexity and wonder of being human, where every day is an opportunity for discovery.

Chapter 5: Modern Genetics and the Debate

Let's embark on an exciting time-travel expedition! Today, we're journeying to the recent past, where modern science unlocked secrets hidden within us and also posed new questions that tug at our ethical fibers. This chapter is a little like uncovering family secrets, with a dash of science fiction thrown in. Ready? Let's go!

Decoding the Human Story: The Human Genome Project

Imagine trying to read a book with three billion letters and no spaces. Overwhelming, right? That's what our DNA looks like! The Human Genome Project was a massive endeavor that aimed to 'read' this incredibly long story written inside each of us.

So, what did this colossal project reveal? Among many findings, one significant realization was that our genes are just part of the story. Despite having a vast genetic code, only a small portion directly dictates our traits. Moreover, these genes don't operate in isolation.

Our environment plays a crucial role in 'switching on' or 'modifying' their effects.

To put it simply: Imagine having a state-of-the-art kitchen (our genetic potential) but not knowing how to use it. Only when someone (our environment) teaches us to cook or provides the right ingredients does the magic happen.

CRISPR: The Scissors of Genetics

Now, if the Human Genome Project was about reading our genetic book, CRISPR technology is akin to having a pair of scissors and some glue, allowing us to edit this book.

CRISPR offers the tantalizing possibility of 'editing' our genes.

Think about it: if a certain gene makes someone more likely to develop a disease, wouldn't it be great to just 'cut it out' and replace it with a healthier version? It's as if you're correcting typos in the story of life.

Treading the Ethical Tightrope

However, as with all great powers, there comes great responsibility. Editing genes isn't just about eliminating diseases. It brings forth ethical dilemmas:

Playing God: If we can edit out perceived 'flaws,' where do we draw the line? Do we risk creating a society that only values certain traits and looks?

Unequal Access: Cutting-edge technologies are often expensive. Could we end up in a world where only the wealthy can afford to 'upgrade' their genes or those of their children?

Ripple Effect: If we change one gene, it might affect others. Are we ready to face unintended consequences?

Furthermore, if we can edit genes to an extent where environment plays little to no role, does it not shift the balance in the nature vs. nurture debate?

The rapid advancements in modern genetics have both enriched and complicated the age-old debate of nature vs. nurture. We've gained the ability to read and potentially edit our genetic story. But, with every chapter we unveil, more questions emerge, many of which venture beyond science into the realm of philosophy and ethics.

As we progress, it becomes essential to balance our thirst for knowledge and capability with wisdom and foresight. The dialogue on nature and nurture isn't just academic; it's deeply personal, affecting every facet of our existence.

Chapter 6: Neuroplasticity: The Brain's Response to Nature and Nurture

Imagine for a moment that your brain is a vast, bustling city. There are well-established routes and highways, but over time, due to construction or traffic patterns, new roads emerge, old ones might close, and some even get a little makeover. This is very similar to how our brain, a beautiful masterpiece of nature, dynamically reshapes itself in response to our experiences.

The Ever-Changing City: Our Brain

Let's start with a magical word – neuroplasticity. It sounds complex, but it simply means our brain's ability to adapt and change. Just like a city might adapt to the changing needs of its inhabitants, our brain reforms its connections in response to our experiences, learning, and even injuries.

Real-life Magic: Stories of Neuroplasticity

Taxi Tales: In London, black cab drivers have to remember thousands of streets and routes for a test called "The Knowledge."

Studies found that, through their intensive training, these drivers have a larger hippocampus (the part of the brain linked to memory) than the average person. It's like a part of their brain city expanded due to heavy traffic!

Musical Minds: Ever heard of the "Mozart effect?" While playing Mozart won't instantly make someone a genius, learning a musical instrument can have profound effects on the brain. Musicians often show enhanced connections in regions linked to hearing and hand-eye coordination.

Bouncing Back: Some people, after a stroke or injury, have lost the function of a part of their body. But with therapy and determination, another part of their brain can sometimes take over this function. It's as if a bridge in our brain city collapsed, but a ferry service started to cover the route.

Nature's Blueprint in the Brain's Adaptability

While our experiences play a significant role in shaping the brain's highways, the foundation of the city, its layout, and initial structures are set by our genes.

For instance, some individuals might have a genetic predisposition that makes their brain more susceptible to anxiety. This doesn't mean they will definitely develop anxiety disorders, but their brain's foundation has a particular design that, when combined with certain life experiences (like trauma), might lead to specific outcomes.

However, on the positive side, understanding these genetic predispositions can guide therapeutic interventions, making them more personalized and effective.

Neuroplasticity is a testament to the beauty of the brain's adaptability, where nature lays down the foundational blueprint and nurture sculpts, refines, and sometimes even redraws parts of it.

It's a dance of resilience and adaptability, showcasing that we aren't just bound by our genes but have the incredible power to learn, adapt, and even overcome. As we journey further into understanding ourselves, remember, our brain, much like a city, is alive, evolving, and always ready for new experiences and challenges.

Chapter 7: The Nature of Intelligence

Imagine intelligence as a shimmering jewel, multi-faceted and radiant. Each facet represents a dimension of our ability to think, reason, learn, and adapt. While we all marvel at its beauty, the quest to fully understand this jewel's origins and properties has been a journey full of questions, discoveries, and debates.

What Do We Mean by "Intelligence"?

When we hear the word 'intelligence,' what comes to mind? For some, it might be excelling in exams. For others, it might be the ability to solve tricky puzzles, or perhaps it's the gift of understanding human emotions. The truth is, intelligence isn't just one thing. It encompasses various abilities, from problem-solving and memory to creativity and emotional insight.

We've tried to measure intelligence, commonly through IQ tests. These tests can give us a snapshot of certain cognitive abilities at a particular time. But remember, it's just one snapshot of a grand, evolving panorama.

Nature's Role in Crafting the Jewel

Now, let's dive into the heart of our exploration: where does intelligence come from? There's considerable evidence to suggest that our genes play a part.

Twin Studies: Research on identical twins (who share all their genes) shows that they often have similar IQ scores, even when raised apart. This similarity hints at a genetic component to intelligence.

Family Matters: It's not uncommon to find that intellectual abilities often run in families, suggesting a genetic link.

The Nurture Nudge in Polishing the Gem

While genes lay the foundation, the environment we grow up in significantly influences the development of our intellectual abilities.

Early Childhood: Studies have shown that children who receive more mental stimulation from their environment (think engaging toys, books, or interaction with caregivers) often display better cognitive abilities.

Education: Access to quality education and learning opportunities can greatly enhance one's intellectual potential. Remember the phrase, "You're a product of your environment"? It's particularly apt here.

Socioeconomic Factors: Children growing up in wealthier households often have access to more resources, better nutrition, and learning opportunities, which can boost intellectual development.

The Delicate Dance of Nature and Nurture

How much of intelligence is due to genes, and how much can be attributed to the environment? This debate has persisted for ages. Some studies suggest that around 50% of intelligence might be genetic. But this doesn't mean the other half is solely due to the environment. Our genes and experiences intermingle in complex ways that we're still trying to understand.

Moreover, what's essential to note is that a genetic predisposition doesn't guarantee an outcome. Think of it like a plant seed (genes) – its potential to grow into a vibrant plant is realized best when given the right soil, water, and sunlight (environment).

The nature of intelligence, much like the jewel we imagined earlier, is intricate and multi-dimensional. As we unravel its mysteries, one fact remains clear: both our genes and our experiences play crucial roles in shaping it. As we continue our journey of understanding, let's celebrate the uniqueness of every individual, recognizing that each person's intelligence is a unique blend, a gem of its own kind.

Chapter 8: The Nurture of Personality

Have you ever attended a family reunion and heard, "Oh, you're just like your Aunt Mildred," or "You have your grandfather's adventurous spirit!"? It's fascinating to think about what makes us who we are. Is it the stories whispered to us as children, the genes passed down through generations, or the unique experiences we collect throughout our lives? In this chapter, we embark on a journey to understand the tapestry of personality.

Unraveling the Threads: What is Personality?

Imagine personality as a colorful quilt. Each patch represents a different trait or behavior. Some patches might symbolize our love for adventure, while others represent our shyness or zest for life. Together, they form a complete picture of who we are.

There have been many theories trying to categorize these traits. Some models talk about 'Big Five' traits: openness, conscientiousness, extra-version, agreeableness, and neuroticism. Others might break it down differently. While these models can provide a framework, remember, everyone's quilt is unique.

The Genetic Thread in Our Quilt

So, how much of our quilt is sewn before we're even born?

Family Ties: Just as physical features run in families, certain personality traits seem to as well. Families might share tendencies toward being outgoing or perhaps having a fiery temper.

Temperament Tales: Babies, even before being influenced much by the world, display distinct temperaments. Some are easy-going, while others are more sensitive to their surroundings. This inherent temperament can be a foundation upon which the rest of the personality quilt is built.

Stitching the Quilt: The Role of Upbringing

While there's a blueprint, the intricate details and designs of our quilt come from our experiences and upbringing.

First Stitches – Family: Early interactions with parents, siblings, and extended family members influence our quilt. The warmth of a family or the coldness of neglect can deeply affect how we view ourselves and the world.

School Days Patchwork: Our school experiences, the friends we make, the teachers we love (or don't), all add layers to our personality. Remember the joy of being praised by a teacher or the hurt from a playground betrayal? These moments shape us.

Culture and Fabric: The culture we're born into provides the fabric upon which our stories are stitched. Values, traditions, and societal norms influence many of our traits and behaviors.

Stories of Nature and Nurture

Anna and Elsa: Let's consider two fictional sisters (inspired by many real-life stories). Both had similar genetic predispositions for being compassionate. Anna grew up in a nurturing environment, where her compassion was encouraged. Elsa, due to certain traumatic experiences, became more reserved, using her compassion in more subtle ways. The same genetic thread, different designs on the quilt.

The Tale of Two Friends: Jack and Oliver, best friends, both loved adventure. Jack's love was born from reading countless books in a restrictive home, craving the outside world. Oliver, on the other hand, inherited his wanderlust from his globe-trotting parents. Same patch, different origins.

The intricate quilt of personality is sewn with threads of genetics, stitched with experiences, and designed with layers of upbringing and culture. As we wrap ourselves in the warmth of understanding, let's appreciate the uniqueness of each quilt and the stories woven into them. Each quilt tells tales of nature's threads and nurture's designs, making every single one a masterpiece of its own.

Chapter 9: Rewiring the Debate

Imagine holding a set of old, tangled headphones. At one end is a jack labeled 'Nature' and at the other end, one labeled 'Nurture'. For years, we've tried to untangle them, trying to see which line plays the loudest music in our lives. But what if the magic happens when both are plugged in together? Let's delve into why it might be time to rewire our old ways of thinking.

Beyond the Old Song and Dance

The age-old debate of nature vs. nurture is much like an intense dance-off, with each side trying to outshine the other. But in reality, life's orchestra plays a duet rather than a solo. Instead of choosing sides, we might gain more by understanding how they dance together.

It's Fluid, Not Fixed: Just like how our tastes in music might change over time, so too does the influence of nature and nurture. At times, genetics might take the lead, while at other moments, our environment shapes our steps.

Seeing Shades: Life isn't black and white. It's a spectrum of colors. Boxing ourselves into a binary debate limits our vision. We need to appreciate the entire rainbow of influences that paint our lives.

A Symphony of Interactions

If we're rewriting our understanding, what might a more nuanced model look like?

The Cascade Effect: Consider a waterfall, where one tier spills into the next. Similarly, our genes might set things in motion (the first tier), but then experiences shape the course and direction of the flow.

Feedback Loops: Think of adjusting the volume of a song. Our environment might turn up certain genetic traits, while dialing down others. Then, these expressed traits further influence how we interact with our surroundings, creating a continuous loop.

A New Tune for Societal Systems

Embracing this intertwined perspective holds powerful implications for various aspects of society:

Education: Recognizing that students are products of both their DNA and their day-to-day experiences encourages a more tailored, flexible approach to teaching. No child is a lost cause, nor is any child solely the product of their home environment.

Mental Health: Understanding the dance between genes and environment helps in creating comprehensive treatments. It also reduces stigma; a person's mental health isn't just 'bad genes' or 'poor upbringing' – it's an intricate interplay.

Public Policy: When we acknowledge that people are shaped by both their biological makeup and their surroundings, we can champion policies that not only provide essential resources but also cultivate environments that nurture positive growth.

As we rewind our old tapes and rewire our perspectives, it's essential to remember that our lives' melodies are a mix. Every chord, every note, is a blend of the songs our genes sing and the tunes the world plays around us.

Let's dance to this new rhythm, celebrating the duet of nature and nurture, and cherishing the unique symphony each one of us represents.

Chapter 10: Future Perspectives

Ever watched a sci-fi movie and wondered, "Is that where we're headed?" From flying cars to robots and space travel, predictions about the future have always stirred our imagination. As we stand at the crossroads of nature and nurture, let's look ahead and envision the future paths this debate might take.

Crystal Ball Gazing: The World of Tomorrow's Research

Science is a journey, always evolving and pushing the boundaries of the known. In the realms of genetics, neuroscience, and psychology, what might the next stops look like?

Mapping More than Genes: We've come a long way since the Human Genome Project. The future might see us not just mapping genes, but understanding the entirety of how they function, interact, and express themselves.

Brain Highways: As neuroscience advances, we could get better blueprints of the brain, understanding not just its structures but its intricate networks. This could shed light on how nature and nurture pave the roads of our minds.

Mind Matters: Psychology might shift from merely observing behaviors to predicting them, using data and sophisticated algorithms. The age-old question of 'why we do what we do' could get some fascinating answers.

Tech: The Game Changer

If there's one thing that's reshaped our world in ways we never imagined, it's technology. How might it rewrite the nature vs. nurture script?

Digital Twins: Imagine having a virtual you. With advancements in AI, we might have digital models of ourselves that simulate how our genes react to different environments. It's like playing a personalized video game of life!

Brain-Computer Interfaces: Devices that connect our brains to computers could offer real-time insights into how our environment shapes our thoughts, decisions, and feelings.

Virtual Environments: As virtual and augmented reality technologies improve, psychologists might study human behavior in entirely controlled, customizable digital worlds, unlocking new dimensions in the nature vs. nurture debate.

Tomorrow's World: The Bigger Picture

The marriage of science and technology promises exciting discoveries. But with power comes responsibility.

Ethical Echoes: From genetic editing to digital surveillance, the future is riddled with ethical questions. What rights do we have over our genes? How much should we alter nature's course? Where do we draw the line between observation and invasion?

Societal Structures: As we gain deeper insights into human behavior, societies might need to reshape education, healthcare, and even legal systems, taking into account the dance of nature and nurture.

Embracing Diversity: Understanding the myriad ways nature and nurture create unique individuals could lead to societies that celebrate differences rather than suppress them.

As the pages of the future unfold, the nature vs. nurture story will gain new chapters, characters, and twists. Whatever the discoveries, they'll serve as a testament to human curiosity and our relentless quest for understanding.

Let's stride into this future, not with trepidation, but with optimism, ensuring that each revelation is used to build a world that's kinder, wiser, and more inclusive.

Chapter 11: Conclusion

Every good song has its ending, allowing us to reflect on its melodies and rhythms. As we near the conclusion of our journey exploring the dance between nature and nurture, let's revisit the beautiful tunes we've encountered and ponder on the symphony they create together.

Echoes from the Past

From the first page to this very moment, our exploration has uncovered:

Historic Harmony: We dove into the age-old debate of nature versus nurture, understanding its roots and significance in the tapestry of human knowledge.

Nature's Notes: Genetics plays its tune in our development, influencing everything from physical characteristics to potential predispositions.

Nurturing Rhythms: The environment, from our early childhood to our cultural surroundings, holds the power to shape our persona, painting atop the canvas laid down by our genes.

Dynamic Duets: Rather than a competition, nature and nurture engage in a harmonious duet, influencing and feeding off each other in intricate ways.

Advancements & Amplifiers: Modern science, from the sequencing of the human genome to the potential of genetic editing, promises to redefine the contours of our debate.

Mysteries of the Mind: Our brain's plasticity showcases the dance of genes and experiences, while our intelligence and personality traits echo the interplay between our biology and biography.

New Perspectives & Playlists: The push is towards transcending binaries and embracing more complex, holistic models of understanding human behavior and development.

Future Frequencies: The horizon holds the promise of even deeper insights, woven together by the threads of research, technology, ethics, and societal evolution.

The Symphony of Self

At the heart of this debate lies a profound truth: we are all unique masterpieces of music, each note a testament to our genetic makeup, and every rhythm a reflection of our life experiences.

Understanding the interplay of nature and nurture isn't just academic; it shapes our worldview. It impacts how we raise our children, how we interact with others, and how we view ourselves.

Knowing that we are products of both our genes and our journeys can instill a sense of empathy, patience, and wonder.

In the Grand Scheme

For society, this debate holds the key to creating nurturing environments, framing policies that respect individual differences, and building inclusive communities that recognize the boundless potential in every individual.

Curtains Close...

The nature vs. nurture debate, like an evergreen song, might never have a definitive ending. And perhaps, that's the beauty of it. It's a melody that invites us to listen, reflect, dance, and discover.

As we turn the page on this chapter and this exploration, let's carry forward the music in our hearts, tuning our ears to the infinite compositions that life offers, and cherishing the unique symphony that each one of us is.

About Freudian Trips

Welcome to Freudian Trips, your dedicated platform for diving deep into the world of psychology. We are more than just a YouTube channel or a book publisher. We are a beacon of enlightenment, making complex psychological concepts accessible and engaging for all.

Our YouTube channel is a rich repository of psychology made simple. We take the profound and often complex ideas from the world of psychology and break them down into digestible, easy-to-understand content. From the foundational theories of Freud to the cognitive insights of Piaget, we cover a broad spectrum of psychological schools and thoughts, making psychology accessible to everyone, regardless of their background or prior knowledge.

As a book publisher, we take the same approach, transforming intricate psychological theories into comprehensible narratives. Our books are not just collections of words, but vessels of wisdom that make psychology approachable and relatable. We believe that psychology should not be confined to academic circles, but should be available to all who seek to understand the human mind and behavior.

At Freudian Trips, we believe in the power of curiosity and the pursuit of knowledge. We are here to stoke the fires of your curiosity, to guide you on your intellectual journey, and to help you navigate the fascinating world of psychology.

If you are someone who is not afraid to question, to explore, and to learn, then you are in the right place. Join us on this journey of exploration, as we make psychology easy to understand, one concept at a time.

Be sure to visit our Youtube channel at: www.freudiantrips.com/youtube

You can also visit us on the web at www.freudiantrips.com

Welcome to The Freudian Trip community. Stay curious. Stay enlightened.

www.ingramcontent.com/pod-product-compliance
Lightning Source LLC
Chambersburg PA
CBHW060902260726
48661CB00008B/3414